CW01369736

1 MONTH OF FREE READING

at
www.ForgottenBooks.com

By purchasing this book you are eligible for one month membership to ForgottenBooks.com, giving you unlimited access to our entire collection of over 1,000,000 titles via our web site and mobile apps.

To claim your free month visit:
www.forgottenbooks.com/free905092

* Offer is valid for 45 days from date of purchase. Terms and conditions apply.

ISBN 978-0-265-88954-1
PIBN 10905092

This book is a reproduction of an important historical work. Forgotten Books uses state-of-the-art technology to digitally reconstruct the work, preserving the original format whilst repairing imperfections present in the aged copy. In rare cases, an imperfection in the original, such as a blemish or missing page, may be replicated in our edition. We do, however, repair the vast majority of imperfections successfully; any imperfections that remain are intentionally left to preserve the state of such historical works.

Forgotten Books is a registered trademark of FB &c Ltd.
Copyright © 2018 FB &c Ltd.
FB &c Ltd, Dalton House, 60 Windsor Avenue, London, SW19 2RR.
Company number 08720141. Registered in England and Wales.

For support please visit www.forgottenbooks.com

INTERIM REPORT

O.R.A. 155-421

Contract # 702930013

A HYDROLOGIC STUDY OF THE SULPHUR ARTESIAN GROUNDWATER SYSTEM
AND ASSOCIATED WATERS AT CHICKASAW NATIONAL RECREATIONAL AREA,
SULPHUR, OKLAHOMA

Prepared By:

Jim F. Harp, Ph.D., Professor
Joakim G. Laguros, Ph.D., Professor
Stephen S. McLin, Ph.D., Assoc. Professor
Leonard B. West, Ph.D., Assoc. Professor

Civil Engineering and Environmental Sciences Dept.
202 West Boyd, Room 334, Carson Engineering Center
The University of Oklahoma, Norman, Oklahoma 73019

Submitted To:

The United State Department of the Interior
National Park Service

Submitted By:

The Bureau of Water Resources Reserach,
Oklahoma Research Administration
The University of Oklahoma
Norman, Oklahoma

June 1984
(Edited, Reviewed, Revised January 1985)

CHAPTER 1

Introduction

Since the late 1970's and early 1980's the National Park Service has been intensely concerned with the negative deprivation impact upon the environment, hydrogeologic and hydraulic conditions in and near the Chickasaw National Recreational Area in Murray County, Oklahoma. The concerns have been many, however, the principal ones are: the cessation of the spring and artesian well flows, the flooding in the stream corridors, the poor quality of the streamflow waters.

Historically, the springs have ceased to flow on several occasions. However, the long dry period from December 1979 until September 1981 was the longest and most severe flow cessation of the springs on record. The major attraction of the entire area is the spring and artesian well flows, some of which are reputed to have medicinal value. These flow cessations have a negative impact upon the value of the Park. More importantly, these flow cessations signal a symptom of very severe problems with water supply and aquifer considerations. As will be shown in this report, the water resources of the area are being over developed and over utilized. A general management strategy scheme is deemed absolutely essential if this area is to remain as a quality visitor attraction site. Time is of the essence.

The park was created by land purchase from the Choctaw and Chickasaw Indian Tribes in 1902 and was named Sulphur Springs

Reservation with an area of about 640 acres. The land was purchased to "embrace all the natural springs in and about the village [Sulphur] and so much of Sulphur Creek, Rock Creek, Buckhorn Creek, and the lands adjacent to said natural springs and creeks as may be deemed necessary by the Secretary of the Interior for the proper utilization and control of said springs and the waters of said creeks....."

In 1906, by join resolution of Congress, the name of the area was changed to Platt National Park. In 1976 the park was enlarged to about 9,500 acres to include the present Arbuckle Reservoir and the name was changed to Chickasaw National Recreational Area. This 1976 addition also included the tract of land upon which the Vendome Well is situated.

The streams, both Travertine and Rock Creek, have flooded several times in the past decade and loss of life has been reported. The character of these streams is noted for rapid rise and fall whenever the Oklahoma weather conditions produce the right precipitation and duration events. The flood ways and fringe delineations have been performed during this first year of this project. Plotted flood ways and flood fringes await close contour maps prepared by others.

The recent opening of the Horsman Well and the drilling of new wells by the City of Sulphur has only added to the current problem of aquifer depressurization. These aspects of the project are being considered as an ongoing concern and will be fully reported by the end of the three year project duration period. Current data is being compiled and the analysis continuing at

this time. The fundamental purpose and intent of this study is to undertake and perform the necessary steps to retain the Park as a Natural Resource and to provide for co-existence with the locale in a manner that will be accommodating and acceptable to all parties concerned. Intensive efforts are being made to assemble accurate information and data upon which to make proper decisions and assessments.

CHAPTER 2

Objectives and Scope of Study

Current project objectives are as described below:

1. To undertake a hydrologic study of the Sulphur artesian ground water system and associated surface waters to develop a clear understanding of the presence, availability, interrelationships and flow characteristics that have existed and are currently in play.

2. To determine, or define, the upper limits of the hydrologic resources that can be developed for mulitpurpose use and still preserve artesian characteristics and aesthetic values of the area.

3. To determine the effects of changes in the hydrologic system on the surface and subsurface flows.

4. To accurately determine the size of the recharge area for the artesian basin, its geologic structure, water level fluctuations from selected observations wells, response of spring levels to pumped wells and to estimate the groundwater artesian discharge rate.

5. To obtain structural data on the dip, strike, joint system and faults in the vicinity of the recharge area in order to determine the pattern of the flow system and variations of permeability and hydraulic conductivity within the system.

6. To assemble data from test holes that have been geophysically logged so as to obtain direct information on the flow system.
7. To assemble available seismic information which will assist in defining the subsurface structure of the recharge area and assess additional covered faults, if any, in the subsurface system.
8. To assemble all available flow measurement data from springs, uncontrolled artesian wells, and production wells, including injection wells, as an analysis base of the flow system study.
9. To perform a flood study whereby the floodways and 100-year flood fringes will be determined in the Travertine Creek and the applicable region of Rock Creek corridors within the confines of the urban sector.
10. To recommend additional data acquisition measurement stations or programs that might be desirable for future planning and management. This element of the study will be determined by the extent of the availability of the data that will be assembled in the early phase of the project.
11. To develop an extensive, comprehensive, resource evaluation document for scientific and lay audiences.

CHAPTER 3

Data and Information Progress Summary

The rainfall records have been assembled from 1917 through early months of 1984. Statistical Subroutine Package SAS was employed on the IBM 3081 and the results presented in Table 1. Monthly averages and yearly averages are presented. Seasonal averages, dry and wet periods, trends, etc. can be easily obtained from the data array and tabulation results.

The water levels in the East and West observation wells for the years 1974-1984, partial, are presented in Tables 2 and 3, respectively. These data indicate the changing status of the groundwater aquifer over this period. In combination with data of water withdrawal from the aquifer, these water levels will be utilized to evaluate the average recharge to the aquifer system. This information will be valuable later in the study when multiple correlation procedures are performed.

The stage of spring flow, an average from the daily recordings, at Buffalo and Antelope Springs are presented in Tables 4 and 5, respectively. These data are converted to average monthly flows, presented in Tables 6 and 7. These were calculated from the physical configurations of the stream channels leading from the springs and the pin readings, depth measurements to a pin set-up by Park personnel

earlier, i.e., a flow vs. depth relationship (rating curve) for the springs has been established.

The existence of these two springs has been documented since the early 1900's. Antelope Spring may be described as a free-flowing spring which issues from a small, protruding bluff, into a shallow, irregular pool of only a hundred or so square feet of surface area. The flow from this spring exits through a rough channel leading from the pool, about 5 feet wide, and then flows about 100 feet to the west before passing under a crosswalk bridge, and emptying into a pond. The flow continues out of this pond, eventually into Travertine Creek.

Buffalo Spring wells up from beneath the surface of the ground into a circular, man-made pool of about 25 feet diameter and 2.5 feet in depth. The water exits this pool by means of a masonry rock-formed channel about 5 feet wide, and then flows over the surface in a small stream, eventually into Travertine Creek. Immediately below Buffalo Spring proper, there is considerable additional water welling up into the creek bed, which increases flow substantially (roughly by a factor of 4).

Measurements of the water elevations stage are made at both springs by Park employees on a irregular basis. These may occur every few days, or only several times a month (or even less frequently), depending upon other workloads at the Park. The measurements are made from steel pins embedded in the pools formed by the springs, prior to their first exit.

These measurements go back to mid-1969 for Antelope Spring, and mid-1972 for Buffalo Spring.

Although the exits from the spring pools are roughly weir-shaped, the exits were not constructed as weirs and thus do not have exact dimensions or smooth surfaces. Therefore, standard weir equations for converting stage to flow are not suitable. Also, the spring depths must reach a certain level before discharge occurs. To complicate matters, the raw data from the earlier analyses by others was not available to this writer, so an attempt was made, with poor success, to correlate past flow data with present data. Agreement was poor at the juncture of these data sets, so that several somewhat arbitrary judgements were made as to which set of data were more applicable.

Essentially, the flow from Buffalo Spring was calculated, assuming the exit from the pool acted as broad, rectangular weir, and flows were calculated for several readings, and the discrepancies were then averaged in the hope of reaching a reasonably accurate value.

For Antelope Spring, the process was somewhat more difficult. There is no easily recognizable weir structure at the pool exit, but the channel immediately below it is reasonable regular, and close approximation to a broad-crested weir exists at the entry to the pond below. Data assembled from the physical configuration were used to establish an approximate flow-vs.-depth relationship.

Basically, for both springs, it may be assumed that,

within certain limits, the flow may be described by the weir equation of the following form:

$$Q = (C)(b)(H)^{1.5},$$

where

Q = flow, in cfs,
C = a constant, dependent upon the channel or weir shape.
= the width of the channel or weir, and
b = the head on the weir at the point in question.

Alternatively, Q and b may be combined into a single term, say "K", reducing the equation to:

$$Q = KH^{1.5}.$$

If the flow at a given depth can be determined by other means, such as critical flow equations, it is then possible to back-calculate "K" for the equation, and the specifics of "C" and "b" are not essential. It must be understood, however, that this derived equation is of limited accuracy, since it does not adequately account for changes in channel section and shape with varying depths. To refine it further, for either spring, would require extensive field studies to accurately measure flows over time, eventually establishing a chart of stage vs. flow. As an alternative, the springs could be restructured to actually exit through weirs of known dimension; this would render invalid all earlier data, however unless such structures were placed far enough downstream to allow the present flow control features to the channel which effect present pin readings to remain unaltered.

One other aspect of these particular springs must be

taken into account, and that is the fact that there must be a finite depth in both springs before flow will occur. This depth will appear in the above equations as a correction to the height (stage) value, as follows:

$$H = P.R. - d$$

where

H = stream depth, as before,
$P.R.$ = the actual pin reading of the depth of the stream, at the point of measurement, and
d = the depth at which flow actually begins over the exit "weir" from the springs.

The equation is now reduced to:

$$Q = L\,(P.R.-d)^{1.5}.$$

To use the equation, it is first necessary to establish the "d" components of the separate equations for each spring. For Buffalo, there is a notation in the October 1978 data files that the depth at which flow began was 1.82 feet; this established "d" for Buffalo. For Antelope, a study was made of flow depths, from the old field notes, and pin readings taken by the Park employees. These, with some variation, established a "d" value of 0.6 feet for Antelope. It should be noted that the pin at Buffalo Spring was discovered to be missing in October of 1981, after an extended dry spell. The pin was replaced, but is not at exactly the same height as the original; based upon field observations and conversations with Park personnel (Mr. Randy Fehr), it appears that the new correction factor, "d", should be 1.97 feet.

Using these "d" values, flow calculations were made us-

ing the older field notes, and data from recent trips to the area. This resulted in the "K" value of 10.44 for Buffalo and 8.69 for Antelope.

For comparison purposes only, and given a 5 feet channel width at Antelope, and a 4.5 feet weir width at Buffalo, the <u>Standard Handbook for Civil Engineers</u>, Frederick S. Merritt, Editor, McGraw-Hill, 1976, on pages 21-71, lists a chart of effective "K" values for broad-crested weirs, taken from King and Brater. Using this chart, "K" values of 11.7, for Antelope, and 10.62, for Buffalo are found. Hence, the value for Buffalo is within 2% of the expected "typical" value, while Antelope is within 25%. Given regularity of the channel shapes of both springs, this is considered to be acceptable, although not ideal, for this portion of this study, at least at this time.

As a result, the following equations have been developed, and have been used in calculating spring flows throughout:

For Antelope Spring;
$$Q = 8.69 \ (P.R.-0.6)^{1.5}.$$

For Buffalo Spring;
$$Q = 10.44 \ (P.R.-1.97)^{1.5}.*$$

*For data prior to October 1981, "d" should be taken as 1.82.

Area well usage data, obtained from the Oklahoma Water Resources Board for the period of 1964 through 1983, partial, are presented in Table 8. The blanks in the data in-

-11-

dicate no records were reported to the State Water Board. The research team will acquire this data from the City.

Establish the Recession Hydrograph of Buffalo and Antelope

The hydrograph is used to predict the flow of water through a drainage basin over time. The descending leg, to the right of the peak, is termed the recession curve, and represents surface water runoff, interflow, and actual groundwater discharge to the stream in question, termed base flow. Over time, the direct runoff ceases first, followed by the interflow, eventually leaving the base flow as the only component of flow. The base flow recession constant, K_r, can be found graphically by plotting a straight line onto the hydrograph where it is adjudged that only base flow is contributing. Given the shapes of the recession portions of the flow curves for these two springs, this method is unsatisfactory. Readings of stage, and consequently flow, are too far apart to yield a useful graph.

Instead, a method proposed by Langbein (W.B. Langbein, "Some Channel Storage and Unit Hydrograph Studies", Transactions American Geophysical Union, Vol. 21, pp. 620-627, 1940), can be used. This method is reported by De Wiest (Roger J.M. DeWiest, Geohydrology, John Wiley and Sons, Inc., 1965, pp. 67-70), and is a graphical technique shown in Figure 14, taken from the referenced book. Using this method, a number of points were plotted from the recession portions of the flow curves, plotting q_n vs q_{n+1}. In this exercise, it was uncertain exactly where the ground-

water portion of the recession curve started, so that the method of selection was somewhat arbitrary. The points selected have quite a scatter to them; however, both sets of points yielded the same recession coefficients, a K_r of 0.7, based upon a "best fit" by eye. A better value will be determined at a later date. Given the relative closeness of the springs to each other, their common source, and near-identical elevation, this is not surprising. It is hoped that future correlations of these springs will be done to confirm or deny this value. This would likely involve several days or weeks of frequent readings, more than has been done to date by Park personnel.

For both springs, K_r = 0.7, as established approximately, to date. Further analysis will be performed in the second and third year of the project.

Other data, such as water quality records, lake-levels, recreational use, etc. is either postponed, or scheduled during the second and third years of this project.

2. Some statistical analysis, as mentioned above, has been performed during this first year of the project. We deem it reasonable to accomplish these final analyses during the third year of the project after maximum data is available. The principal item of interest is the low flow and no flow data and their correlation with water usage and hydraulic refinement. All of this data is not actually available, but is being sought. For instance, the City of Sulphur water use which was unregulated for 1982 and 1983 but is

determinable from City Water Department sources with some difficulty.

3. The applicability of various hydrologic and hydraulic models is progressing. At this time, the use of HEC-1 and HEC-2 has been performed and further evaluation will extend into the second and third years of the study. It has been tentatively established that the Stanford Watershed model and other dynamic models are not applicable to the Sulphur area because they represent a time dependent continuous model, as compared to the fully valid single event type models such as HEC-1 and HEC-2.

4. The investigation of the slip, fracture, joint and other geological structure analyses has begun with on site field investigations. The research team feels that aerial photographic stereo pairs will be essential for the intensive effort deemed necessary in the latter phases of this project. Aerial photographs have been requested from the team who has made photogrammetric analysis, not yet completed. Other aerial photographs are available at the Oklahoma Geological Survey, here on the Oklahoma University campus, and the analysis continues at this time.

During Phase II of the project, Task 5 of the original proposal will be fully and completely addressed. Initial efforts in the determination of the fracture pattern throughout the recharge area to the Sulphur artesian groundwater system will focus on a data-base acquisition,

consisting of structural information available from geological reports and geophysical logs. It is anticipated, however, that only the most recent aerial photographs will provide quality indicators to the geographical distribution of complex fracture sets indicated to the geographical distribution of complex fracture sets indicated by parallel or subparallel lineaments on these photos. This technique involves the determination of the number and orientation of fracture sets within a number of small subareas of the recharge area, and a statistical comparison of this data among these various subareas, or stations. These stations will be intentionally located along projections of the Sulphur syncline fold axis and fault, along portions of the fold limbs, and other critical positions in order to establish a relationship between the fracture sets and major structural features within and surrounding the recharge area.

The strike and dip of a large sample of fractures will be plotted and contoured on a point diagram after a sterographic projection corrects for changes in strike and dip resulting from block rotations (if actually present). The centers of concentrated fracture sets then indicates attitudes of fracture sets.

Several fracture sets frequently appear together, especially in areas of multiple deformation. Hence, field verification of aerial photo interpretations will be required. Some useful observations will include actual strike and

dip data collection, fracture fillings, features on fracture surfaces, and fracture spacing.

Finally, after a map of structural features is constructed, an overlay will be made showing locations of flowing or production wells and springs. Historically reported springs which are no longer active will also be sited. A comparison of spring or well yields located near fracture sets will be made to well yields located away from fracture sets. This procedure may shed important insight of the influence of subsurface fluid movement from the recharge area to these discharge points. If sufficient information exists, then a piezometric head contour map will be prepared. The final objective of Task 5 is to establish the relative importance of fracture dominated fluid migration from the recharge area to discharge areas within the park boundary.

5. The major accomplishments of this first year have been the near completion of the flood study made on Travertine and Rock Creeks through the Park riverine corridors. Since the maps upon which the flood ways and flood fringes are not yet available, but are expected by mid-summer 1984, only a review of the flood study will be included at this time. The flood study will be a separate document to be published in late summer or early fall of 1984. Summarily, the flows have been determined from the HEC-1 computer program to be:

	Sta/Q(cfs)	10yr	50yr	100yr	500yr
Travertine Nature Center	0 + 00	874	1,121	1,309	1,695
Travertine Confluence	62 + 00	3,380	4,301	5,116	6,644
Foot Bridge	75 + 66	4,123	5,255	6,213	8,063
Rock Creek Confluence	76 + 40	7,805	10,183	12,265	16,301

Complete output including cross-section, drainage areas, backwater data, and floodway encroachment station will be included in the Flood Study document proper anticipated to be available soon, whenever the National Park Services provide the close contour maps, estimated to be available by July 1, 1984.

917		.62	.17	4.68	2.28	3.04	3.12	2.15	3.09	T	1.72	T	
918	.62								4.15	4.37	2.56	3.92	
919	1.21	2.39	2.18	3.40	5.35	3.67	3.81	7.25	.79	8.04	3.86	1.10	43.04
920	2.65	.14	1.45	3.22	6.08	4.88	1.64	2.15	5.38	5.64	1.18	1.48	35.89
921	1.66	2.17	4.91	4.27	2.80	8.00	3.90	2.83	.78	.05	1.11	1.06	33.54
922	1.40	1.36	3.09	8.31	6.08	1.11	2.58	.79	.04	2.33	4.50	.40	31.99
923	2.51	1.28	1.80	5.30	8.88	3.87	T	2.51	6.54	10.75	2.81	4.26	50.51
924	.89	.79	3.42	5.88	3.10	3.63	.35	.72	1.98	.14	.49	2.85	24.24
925	.99	.78	.08	2.19	4.83	.02	2.95	1.32	5.08	3.63	1.48	.32	23.67
926	3.47	.72	2.56	2.56	4.40	3.17	6.19	8.28	5.84	5.39	.66	3.31	46.55
927	4.78	2.97	1.95	9.99	.76	6.44	7.70	2.93	5.00	3.19	.77	3.65	50.13
928	3.30	2.99	1.10	7.86	3.20	9.77	4.97	3.06	.76	3.55	2.68	1.74	44.98
929	2.04	1.22	3.52	1.87	10.46	2.49	1.84	.14	5.05	6.76	2.36	2.78	40.53
930	2.10	2.76	1.23	3.73	7.99	2.57	1.18	3.42	1.73	2.47	4.02	2.34	35.54
931	.35	7.20	4.30	2.02	1.83	1.63	2.95	.72	.14	6.22	6.03	1.44	34.83
931	7.33	3.41	1.73	3.71	2.11	4.06	1.48	1.76	.44	2.54	.22	9.30	38.09
933	2.50	2.02	4.40	2.25	13.78	.54	2.22	5.95	4.96	.46	1.82	1.44	42.34
1934	2.40	2.72	3.80	2.97	3.00	2.81	.00	.87	6.38	1.06	5.98	.48	32.47
1935	1.89	.85	5.38	4.31	12.54	7.48	2.34	4.14	5.45	3.84	3.09	3.05	54.36
1936	.43	.40	1.61	1.07	8.38	.60	2.52	T	12.68	3.56	.31	1.62	33.17
1937	2.47	.18	4.06	3.68	2.22	2.79	4.63	4.38	.05	3.27	2.10	2.93	32.76
1938	2.60	9.55	4.61	2.35	6.05	4.83	2.57	1.43	2.44	.50	2.80	.68	40.41
1939	3.07	1.73	1.96	2.78	2.70	4.35	1.26	2.67	.59	2.91	2.28	1.20	27.50
1940	.45	2.87	T	6.82	9.33	6.58	6.83	2.47	.22	3.28	6.67	2.87	48.39
1941	2.88	3.18	.47	6.89	4.62	7.71	1.06	6.63	4.02	14.89	1.59	1.58	55.52
1942	.46	2.30	1.70	10.60	4.28	8.18	.77	4.76	3.68	5.41	2.24	2.53	46.93
1943	.15	.74	3.14	4.23	7.81	2.73	1.21	T	1.21	1.63	.12	3.87	26.84
1944	2.78	4.76	4.63	2.75	5.86	2.54	3.12	2.77	.85	4.44	3.82	2.65	38.97
1945	1.70	4.44	9.76	8.61	1.41	10.99	4.58	6.13	11.13	.78	.72	.08	60.33
1946	6.42	3.66	4.01	2.83	5.66	3.43	.84	6.17	2.72	.14	6.80	8.42	51.64
1947	.32	.34	1.21	9.04	8.04	4.28	1.57	1.13	3.33	1.61	2.87	2.70	39.34
1948	.98	4.19	2.51	.70	7.59	6.64	3.62	1.32	.12	.78	.48	1.27	30.20
1949	5.97	2.69	3.38	2.18	6.14	4.90	.58	2.75	6.28	4.54	.00	1.48	40.84
1950	3.02	1.86	.35	2.03	6.44	3.39	5.91	7.94	2.48	.71	.22	.10	34.45

continued)

Table 1 (continued)

RAINFALL RECORD

Chickasaw National Recreation Area
Sulphur, Okla.

For the Years 1917-1984

Year	Jan	Feb	Mar	Apr	May	Jun	Jul	Aug	Sep	Oct	Nov	Dec	Total
1951	1.14	3.61	1.35	1.43	5.88	6.87	5.54	2.92	1.78	3.15	1.84	.31	35.82
1952	.31	1.23	3.33	4.72	4.86	.50	2.85	.90	.10	.03	4.47	1.45	24.75
1953	.55	.97	3.14	4.36	4.51	2.28	7.72	1.44	1.16	5.15	2.22	1.05	34.55
1954	1.49	.55	.93	5.45	8.21	4.34	.15	.83	.36	5.72	.36	3.08	31.47
1955	1.47	2.18	2.45	1.81	4.77	1.79	2.09	1.08	7.37	.21	.02	.40	25.64
1956	1.47	2.14	.39	2.89	4.81	1.55	1.43	1.39	.04	3.73	3.93	3.15	26.92
1957	2.13	2.09	4.37	9.14	13.24	6.09	2.34	1.68	11.41	2.25	4.86	1.46	61.06
1958	2.44	.51	4.15	2.06	4.85	3.52	4.30	1.91	.78	.33	1.94	.86	27.65
1959	.45	.79	3.36	2.74	4.45	2.87	3.04	2.07	1.81	8.96	2.50	2.77	35.81
1960	2.74	3.08	2.10	2.76	6.89	1.84	3.59	3.16	5.33	4.38	.21	4.95	41.11
1961	.10	2.00	3.62	.92	3.75	3.51	3.70	.64	6.87	3.72	3.20	1.74	33.67
1962	.34	.35	2.55	2.97	1.91	8.91	4.13	1.19	4.69	4.85	2.75	1.79	36.43
1963	.07	.04	5.02	3.22	.81	.56	2.28	1.99	1.77	.04	2.29	1.45	19.54
1964	.96	1.59	2.89	1.79	8.52	.98	.70	4.98	6.33	.47	6.70	.72	36.63
1965	2.14	1.55	1.65	2.60	5.90	1.69	1.58	2.48	2.04	1.95	.83	1.32	25.73
1966	1.26	1.87	1.26	5.09	.71	2.18	4.65	5.36	2.04	1.86	.94	.69	27.91
1967	.34	.80	1.05	8.17	6.68	6.73	3.89	.20	5.81	5.37	.79	2.13	41.96
1968	4.13	1.27	3.82	2.35	9.47	4.02	5.75	.97	4.43	2.87	5.73	2.04	46.85
1969	1.25	2.97	2.55	5.21	5.05	3.30	1.74	1.61	2.65	7.92	.51	3.47	38.23
1970	.18	1.74	3.90	4.97	2.26	4.82	1.68	.80	11.73	16.43	.79	.77	50.07
1971	1.89	1.15	.68	2.76	2.83	2.63	4.05	4.92	2.79	6.62	.57	5.46	36.35
1972	.36	.56	.61	3.95	4.05	2.39	1.41	3.56	2.52	7.69	3.87	.88	31.85
1973	3.01	3.00	6.44	6.04	2.72	8.12	3.94	.99	10.90	4.89	8.02	.86	58.93
1974	.90	1.76	1.45	4.91	3.64	3.47	2.10	5.49	8.65	7.10	2.48	1.53	43.48
1975	1.98	4.26	4.71	3.05	6.85	2.43	4.73	1.98	3.55	1.24	2.05	1.50	38.33
1976	.07	.29	3.70	3.88	7.78	1.33	1.90	1.18	1.36	4.71	.90	1.57	28.67
1977	1.91	1.39	6.39	3.51	5.29	2.48	2.04	2.83	1.89	1.73	1.29	.34	31.09
1978	1.15	2.82	3.69	3.00	8.12	3.33	2.13	1.95	2.37	1.31	4.46	.58	34.91
1979	2.33	1.31	4.56	3.68	4.75	6.83	.80	2.72	2.63	3.18	.95	1.69	35.43
1980	1.44	1.26	.91	.56	9.58	2.50	.18	.00	10.73	1.74	1.79	2.29	32.98
1981	.29	2.61	2.52	2.29	8.51	3.46	3.95	1.81	2.10	16.26	1.71	.26	45.77
1982	1.75	2.29	1.60	1.04	15.71	6.21	2.95	1.04	1.02	2.84	3.43	1.77	41.65
1983	2.57	3.70	2.95	2.95	10.67	3.64	.90	3.28	2.11	4.53	2.61	.57	40.48
1984*	.98	1.98	-	-	-	-	-	-	-	-	-	-	(2.96)
Avg.	2.08	2.10	2.78	3.94	5.66	3.95	2.78	2.61	3.65	3.97	2.40	1.98	37.90

Table 2

GROUND WATER LEVELS
EAST OBSERVATION WELL

Chickasaw National Recreation Area
Sulphur, Okla.
(Approximate Surface Elevation = 1150')

Depth to water, below land surface datum
(Beginning of Month Readings)

Year	Jan	Feb	Mar	Apr	May	June	July	Aug	Sept	Oct	Nov	Dec
1974*	38.5	44.5	47.0	49.5	51.0	51.0	54.0	58.0	61.0	59.0	56.0	44.0
1975*	49.0	51.0	42.0	39.0	41.0	40.5	44.5	50.0	56.0	59.5	62.0	64.0
1976*	66.0	67.5	69.5	68.0	67.0	64.5	65.0	63.5	69.5	71.0	71.0	73.0
1977*	74.5	76.5	75.5	67.0	64.0+	63.0	64.5	66.0	57.0	59.0	60.5	61.5
1978*	63.0	66.0	76.0	73.5	71.0	67.5	61.0	64.5	66.0	68.5	71.0	73.0
1979*	75.0	76.0	77.0	71.0	67.0	63.5	62.0	63.5	66.5	68.5	70.0	71.5
1980*	--	--	76.5	76.5	81.0	82.5	81.5	83.5	87.0	86.0	86.0	87.0
1981	86.5*	88.0*	89.0*	86.5*	86.5*	85.5*	75.5*	76.8	78.1	80.2	59.2	58.1
1982	61.6	59.8	57.2+	58.7	60.5	48.0+	44.2	47.9	53.2	57.0	61.2	64.8
1983	66.8	66.7+	62.0	61.5	60.8	45.2	49.7	56.1	60.9	63.1	65.5	66.3
1984	70.0	70.9	71.9+									

NOTE: Flow at both Antelope and Buffalo Springs cease when this well reaches a depth of between 68 and 72 feet below the ground surface.

* Data taken from graph - original data unavailable

\+ Denotes questionable value

Table 3

GROUND WATER LEVELS
WEST OBSERVATION WELL

Chickasaw National Recreation Area
Sulphur, Okla.
(Approximate Surface Elevation = 1080')

Depth to water, below land surface datum
(Beginning of Month Readings)

Year												
1974*	23.0	24.5	23.5	24.0	23.0	25.0	26.5	27.0	28.5	25.5	23.0	23.0
1975*	24.5	23.5	23.0	20.5	21.0	20.5	23.5	23.5	26.0	27.0	27.0	27.0
1976*	16.5	27.5	29.0	28.0	28.0	26.5	29.0	29.0	30.5	30.5	31.0	31.0
1977*	31.5	31.0	30.0	29.0	29.0	28.0	28.5	29.5	30.5	30.5	30.5	30.5
1978*	30.5	30.5	29.5	29.5	29.5	29.0	29.0	29.0	29.0	29.0	29.0	29.0
1979*	29.5	29.5	29.0	29.0	28.5	25.0	26.0	31.0	31.0	31.5	32.5	33.0
1980*	--	--	32.0	32.0	33.0	33.0	33.0	34.5	34.5	33.0	34.0	33.5
1981	32.0*	32.5*	32.5*	31.5*	32.5*	30.8	31.8	33.0	33.0	33.5	28.8	28.6
1982	32.2	28.5	28.3+	29.4	30.2	26.3	27.3	29.1	29.8	29.0	29.5	30.5
1983	--	28.7+	28.3	28.1	28.6	27.2	28.1	29.2	30.3	30.6	29.3	29.5
1984	29.2	29.2										

* Data taken from graph - original data unavailable

+ Denotes questionable data

Table 4

STAGE DATA
BUFFALO SPRINGS

Chickasaw National Recreation Area
Sulphur, Okla.
(Approximate Surface Elevation = 1078')

Year												
1972	-	-	-	-	-	1.95	1.91	1.99	1.49	1.06	2.00	-
1973	2.08	2.10	2.15	2.18	2.16	2.13	2.10	2.08	2.08	2.12	2.18	2.17
1974	2.15	2.12	2.08	2.08	2.08	2.09	2.02	2.01	2.03	2.07	2.13	2.11
1975	2.10	2.13	2.15	2.13	2.13	2.10	1.00	2.05	2.05	2.01	1.99	2.00
1976	1.97	1.94	1.96	1.96	1.99	1.98	1.94	1.90	1.86	1.56	1.33	0.52
1977	Dry	Dry	Dry	1.99	2.00	1.99	1.97	1.95	1.92	1.82	1.54	1.00+
1978	0.46+	Dry	Dry	1.79	1.93	2.01	2.01	1.95	1.90	1.76	1.36	Dry
1979	Dry	Dry	1.93	1.97	2.00	2.01	1.02	1.07	1.92	1.90	1.59	0.93
1980	Dry	Dry	Dry	Dry	Dry	Dry	Dry	Dry	Dry	Dry	Dry	Dry
1981	Dry	Dry	Dry	Dry	Dry	Dry	Dry	Dry	Dry	2.20	2.25	2.22
1982	2.23	2.22+	2.22+	2.22	2.26+	2.26	2.21	2.21	--	2.15+	2.13+	2.14+
1983	2.16	2.29	2.20	2.20	2.28	2.23	2.21	2.20	2.16+	2.70	2.10	1.49
1984	1.75	1.41										

+ Denotes questionable data

Monthly average as derived from field data collected during the month.

Table 5

STAGE DATA
ANTELOPE SPRINGS

Chickasaw National Recreation Area
Sulphur, Okla.
(Approximate Surface Elevation = 1080')

Year												
1969	-	-	-	-	-	0.95	0.92	0.88	0.87	0.88	0.88	0.89+
1970	0.87	0.86	0.89	0.92	0.93	0.86	0.83	0.79	0.82	0.96	0.94	0.94
1971	0.91	0.89	0.87	0.83	0.80+	0.75	0.69	0.70	0.64	0.70	0.70	0.87
1972	0.88	0.87	0.85+	0.84	0.84	0.78	0.73	0.68	0.62+	0.54	0.87	0.89
1973	0.90	0.97	1.05	1.08	1.07	1.05	1.01	0.99	1.01	1.02	1.11	1.08
1974	1.06	1.05	1.05	1.06	1.03+	1.01	0.96	0.91	0.92+	1.04	1.03	1.02
1975	1.05	1.11	1.20	1.23	1.23	1.19	1.00	0.98	0.95	0.95	0.90	0.88
1976	0.76	0.84	0.86	0.86	0.96	0.94	0.84	0.80	0.73	0.65	Dry	Dry
1977	Dry	Dry	0.84	0.92	0.91	0.90	0.87	0.83	0.80	0.77	0.70	Dry
1978	Dry	Dry	Dry	0.71	0.80	0.88	0.84	0.81	0.76	0.68	0.65	Dry
1979	Dry	Dry	0.78	0.78	0.88	0.90	0.84	0.80	0.75	0.74	0.68+	Dry
1980	Dry	Dry	Dry	Dry	Dry	Dry	Dry	Dry	Dry	Dry	Dry	Dry
1981	Dry	Dry	Dry	Dry	Dry	Dry	Dry	Dry	Dry	0.95	0.99	0.99+
1982	0.94+	0.96+	0.98+	-	0.94	1.02	0.93	1.91+	1.74+	-	0.83+	0.83+
1983	0.79	0.87	0.90	0.88	1.00	0.94	0.92	0.84+	0.87+	0.81	0.79	0.78
1984	0.64	0.40										

+ Denotes approximate value

Monthly average as derived from field data collected during the month

Table 6

SPRING FLOW IN CUBIC FEET PER SECOND
BUFFALO SPRINGS

Chickasaw National Recreation Area
Sulphur, Okla.

For the Years 1972 - 1983

Year													
1972	-	-	-	-	-	0.49	0.28	0.73	0.00	0.00	0.80	0.00
1973	1.38	1.55	1.98	2.26	2.07	1.80	1.55	1.38	1.38	1.72	2.26	2.16
1974	1.98	1.72	1.38	1.38	1.38	1.46	0.93	0.86	1.00	1.31	1.80	1.63
1975	1.55	1.80	1.98	1.80	1.80	1.55	0.73	1.15	1.15	0.86	0.73	0.80
1976	0.61	0.43	0.55	0.55	0.73	0.67	0.43	0.24	0.08	0.00	0.00	0.00
1977	0.00	0.00	0.00	0.73	0.80	0.73	0.61	0.49	0.33	0.00	0.00	0.00
1978	0.00	0.00	0.00	0.00	0.38	0.86	0.86	0.49	0.24	0.00	0.00	0.00
1979	0.00	0.00	0.38	0.61	0.80	0.86	0.33	0.55	0.33	0.24	0.00	0.00
1980	0.00	0.00	0.00	0.00	0.00	0.00	0.00	0.00	0.00	0.00	0.00	0.00
1981	0.00	0.00	0.00	0.00	0.00	0.00	0.00	0.00	0.00	1.15+	1.55	1.30
1982	1.38	1.30	1.30	1.30	1.63	1.63	1.23	1.23	-	0.80	0.67	0.73
1983	0.86	1.89	1.15	1.15	1.80	1.38	1.23	1.15	0.86	6.51*	0.49	0.0
1984	0.00	0.00	-									

All flows calculated using the equation: (for data thru October of 1981)

$$Q = 10.44(P.R.-1.82)^{1.5}$$

After October of 1981, the equation was changed to:

$$Q = 10.44(P.R.-1.97)^{1.5}$$

where: Q = flow in cfs,
P.R. = pin reading (stage of spring, in feet

* Reading of doubtful accuracy

+ Pin reset, approx. 0.15' lower than original setting

Monthly mean flow as derived from data of Table 4.

Table 7

SPRING FLOW IN CUBIC FEET PER SECOND
ANTELOPE SPRINGS

Chickasaw National Recreation Area
Sulphur, Okla.

For the Years 1969 - 1983

Year	Jan	Feb	Mar	Apr	May	Jun	Jul	Aug	Sep	Oct	Nov	Dec
1969	-	-	-	-	-	1.80	1.57	1.29	1.22	1.29	1.29	1.36
1970	1.22	1.15	1.36	1.57	1.65	1.15	0.96	0.72	0.90	1.88	1.72	1.72
1971	1.50	1.36	1.22	0.96	0.78	0.50	0.23	0.27	0.07	0.27	0.27	1.22
1972	1.29	1.22	1.09	1.02	1.02	0.66	0.41	0.20	0.02	0.00	1.22	1.36
1973	1.43	1.96	2.62	2.89	2.80	2.62	2.28	2.12	2.28	2.37	3.17	2.89
1974	2.71	2.62	2.62	2.71	2.45	2.28	1.88	1.50	1.50	2.54	2.45	2.37
1975	2.62	3.17	4.04	4.35	4.35	3.94	2.20	2.04	1.80	1.80	1.43	1.29
1976	0.56	1.02	1.15	1.15	1.88	1.72	1.02	0.78	0.41	0.10	0.00	0.00
1977	0.00	0.00	1.02	1.57	1.50	1.43	1.22	0.96	0.78	0.61	0.27	0.00
1978	0.00	0.00	0.00	0.32	0.78	1.29	1.02	0.84	0.56	0.20	0.10	0.00
1979	0.00	0.00	0.66	0.66	1.29	1.43	1.02	0.89	0.50	0.46	0.20	0.00
1980	0.00	0.00	0.00	0.00	0.00	0.00	0.00	0.00	0.00	0.00	0.00	0.00
1981	0.00	0.00	0.00	0.00	0.00	0.00	0.00	0.00	0.00	1.80	2.12	2.12
1982	1.72	1.88	2.04	-	1.72	2.37	1.65	13.03*	10.58*	-	0.96	0.96
1983	0.72	1.22	1.43	1.29	2.20	1.72	1.57	1.02	1.22	0.84	0.72	0.66
1984	0.07	0.00										

All Flows calculated using the equation:

$$Q = 8.69(P.R. - 0.6)^{1.5}$$

where Q = flow in cfs
$P.R.$ = pin reading (stage of spring, in feet)

* Reading of doubtful accuracy

Monthly mean flow as derived from data of Table 5.

Table 8

WELL PUMPAGE DATA
AREA WELL FIELDS

Chickasaw National Recreation Area
Sulphur, Okla.

Year	O.G. & E Wells	City of Sulphur	Winrock Ranch
1964	--	--	169.44
1965	348.66	240.14	102.97
1966	296.52	364.95	--
1967	291.31	300.11	--
1968	302.39	210.83	162.93
1969	328.46	211.80	173.68
1970	440.55	228.10	173.68
1971	334.98	235.26	217.34
1972	348.66	271.43	--
1973	--	361.56	--
1974	--	340.45	--
1975	--	274.71	--
1976	--	298.73	--
1977	--	336.00	--
1978	--	--	--
1979	--	--	--
1980	472.39	--	--
1981	466.97	224.82	--
1982	480.88	--	--

(All data in million gallons)